Coffee with Barbie® Doll

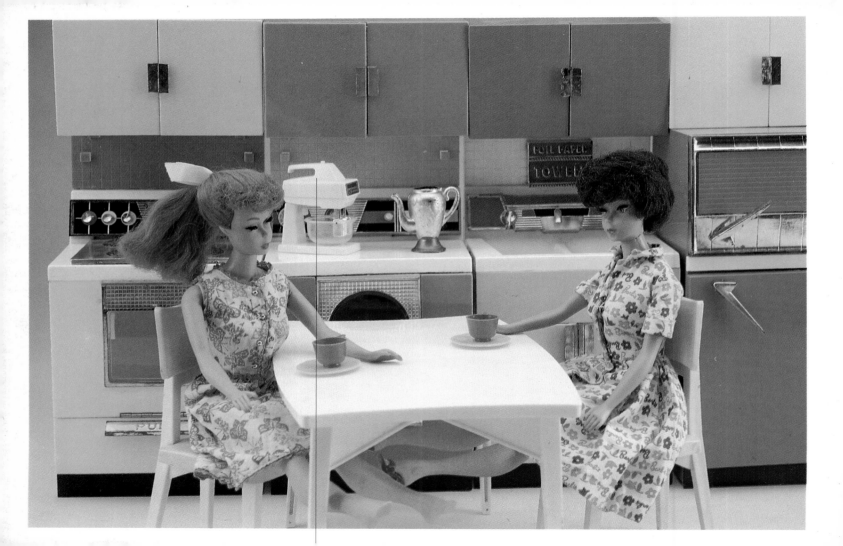

4880 Lower Valley Road, Atglen, PA 19310

Sandra "Johnsie" Bryan

Coffee with Barbie ® *Doll*

Dedication

This book is dedicated to John "Skip" Bryan, my husband. He is the best Ken a girl could ever wish for.

Book design by Blair R.C. Loughrey

ISBN: 0-7643-0412-7
Printed in China
1 2 3 4

Published by Schiffer Publishing Ltd.
4880 Lower Valley Road
Atglen, PA 19310
Phone: (610) 593-1777; Fax: (610) 593-2002
E-mail: Schifferbk@aol.com
Please write for a free catalog.
This book may be purchased from the publisher.
Please include $3.95 for shipping.

Try your bookstore first.

We are interested in hearing from authors with book ideas on related subjects.

I want to say a big "Thank You" to:

My late mother Mavis and my daughter Jennifer who gave me the beginnings of my collection. To my son John who, as a kid, went with me on all those doll hunts.

To Sibyl DeWein who co-authored the first book on Barbie® dolls in 1977.

To Ruth Cronk in New York and the late Dori O'Melia of Arizona who published the first newsletters for collectors in the late 1970s.

To Joan Amundsen and Leonard Planes who compiled the first vintage Barbie® doll clothing guides (serial numbers 900 and 1600) in 1982 and 1984.

To the late Linda Pilkenton for her first Skipper® doll clothing guide in 1990.

To Sarah Sink Eames for the first all-color clothing guides for the entire Barbie® doll line published in 1990 and 1997.

To my first Barbie® friends: Irene Norris, Nora Celio, the late Virginia Sherrod, Margie Englund Kukuk, and Peggy Dyas. Each of you helped me at a special time.

To Carol Spencer for designing so many of my favorite doll outfits, especially from the Eyelash Era, 1967-1972.

And to Mattel, Inc., which continues to make the doll that is the most fun of all!

This book is designed to represent Barbie® doll play scenes as imagined by the author. Neither the outfits nor the structures are factory complete, but rather they are combined in unique and imaginative ways. The values of the clothes, structures, and accessories are given in each caption. Since they are pictured many times, the doll values are listed in the Value Guide at the end of the book. All values are for items which have been played with but nevertheless are in very good to excellent condition. This is the condition most collectibles are in nowadays. You should double the price of items which are perfect and in their original boxes because these are extremely rare finds. Halve the price of items which are soiled, repaired in any way, or missing small parts. Any other condition is not collector quality and should be valued at garage sale prices. These values are the author's idea of what a fair market value might be at the time of publication. Please keep in mind that dealers pay half of what an item might sell for and markets do vary. Age alone is not an indicator of price. Rarity and demand (two popular factors among contemporary collectors) are much stronger indicators for determining current prices of older dolls and accessories.

Most of the dolls pictured are from 1959 to 1984, the 25th Anniversary of the Barbie® doll. This is as good a stopping point as any for collectors, but odds are I won't!

Contents

We have settled in on the sofa, side by side, happy and joking and full after a great little dinner for two. We are all comfy in my tiny but adorable apartment and he is about to make his first move when ... darn it, we are interrupted by Mom! It seems that my real dinner is ready and it's time for me to set the table. I hope she didn't think anything was unusual. Was I blushing? Caught in the act! No, it's not what you're thinking. It's just that I am too old to be playing with dolls!

This scene has, no doubt, been repeated thousands of times since 1959 when a new fashion doll named Barbie® doll hit the toy store shelves. She was not the first doll with a figure and the ability to wear high heels. That honor may go to the Mme. Alexander doll, "Cissy," who first appeared in 1954. Nevertheless, the Barbie® doll stood out with her stunning looks and extensive wardrobe. With success, her lineup of cars and houses soon followed. The introduction of boyfriend Ken® doll in 1961 created a furor because little girls could really play grownup. So a little sister, cousin, and tiny twin siblings, along with countless friends, were created to satisfy every parents' idea of what dolls were appropriate for their particular child. None of this mattered much. Barbie® doll remained the ever-increasing favorite of little girls for decades. She is the most astounding marketing success the world has ever seen, right up there with Ford and IBM. More than a billion Barbie® dolls have been sold, and she currently makes Mattel, Inc. well over a billion dollars a year.

Back then we imagined what life would be like for us when we were all grownup. Our play created a reality that exactly suited each one of us as we tried out different careers and life events. So come along with me as we revisit our favorite playpal. Settle in with a cup of coffee. Get ready to turn the pages and have some fun!

Chapter One

She arrived in a box in 1959. It is very attractive with drawings of many of the outfits, available separately of course, as well as an indication of what color hair and hairstyle the doll inside had. With a hanky the box was transformed into her bed, or pushed against the wall it was her sofa. These boxes were a little larger than the doll, so they could hold her stand and a modest wardrobe. This is probably why so many have survived.

The first house — really an apartment — came in 1962. It is made completely of cardboard, including the furniture, and fulfilled every girl's fantasy of independence. There is a living room with a stereo/TV which flows into the bedroom. Presumably she led the best life of all — there is no kitchen, so she ate out every meal!

By 1984, ten more houses had been added, one three-story townhouse, and at least two houses which were marketed exclusively by a department store (Dept. Store Special or DSS). In addition, there were vinyl cases with fold-down beds, dressing tables, or molded in furniture units. Her portfolio of real estate is impressive! If you are the lucky owner of one of these, be careful. The cardboard ones are fragile, and bugs just love them. The vinyl ones split easily, and the dolls will leave a melt mark on the early plastic furniture wherever their unprotected limbs rest for a while.

In 1957, Mattel manufactured wooden furniture roughly the size of Vogue's Ginny doll. These charming modern pieces include living and bedroom suites and, though not intended to, fit the early Barbie® dolls very well since their legs are straight and thus closer to the ground. Later Mattel made plastic furniture that was sold separately. These are the most fun because we can create any rooms we want!

Suzy Goose Toys made licensed furniture in the 1960s which includes beds, dressers, armoires, and a piano. Deluxe Reading created the essential four-piece kitchen in 1963 that has become the favorite of collectors for its colors and modern design lines. Many other companies such as Ideal and Kenner also made great houses, and this is usually how we played, mixing them all up our own way. But I bet your apartment was a lot like mine!

When we girls get together over coffee, it's usually in the kitchen. We wear a casual house dress and talk while cooking and cleaning. Our kitchen has the latest of everything in the most modern colors. In the '60s our favorites were turquoise and pink, and our dresses matched our kitchen cabinets!

Dolls: 1963 Red Ponytail in 1964 Brunch Time, $75; 1963 Brunette Bubble Cut in 1964 Barbie Learns to Cook, $75. Scene: 1963 Kitchen by Deluxe Reading Corp., $175. 1987 Action Accents mixer by Mattel, $5.

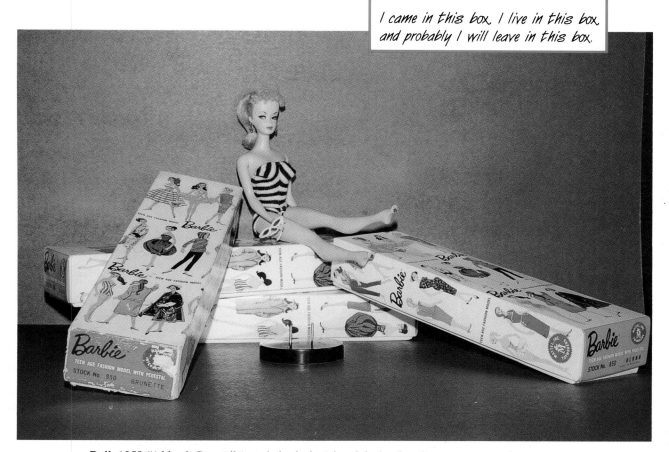

I came in this box. I live in this box and probably I will leave in this box.

Doll: 1959 #1 blonde Ponytail (note holes in feet) in original swimsuit.
Scene: Left box, 1959, $500; other boxes — Mattel Festival gifts for convention goers, Orlando, 1994, $1,000 each (with dolls).

11

Finally! It's so much fun to have my own apartment. I can do anything I want, even have chocolate cake for breakfast!

Doll: 1959 #1 blonde Ponytail in 1959 Nightie Negligee, $50. Scene: 1961 Suzy Goose canopy bed, $25. Pillow, cake, and tray from author's collection.

Doll: 1960 #3 brunette Ponytail. Clothes: 1962 Pak lingerie and dresses, $20 each. Scene: 1962 Suzy Goose Wardrobe, $30; 1958 Wood chair by Mattel, $25; plant from author's collection.

What should I wear today? Will it be the blue one, the red one, or the green one?

13

Doll: 1960 #3 brunette Ponytail in 1959 Sweet Dreams, $50. Scene: 1958 wooden furniture by Mattel, $75 for all, and Mattel accessories, $5 each. Teddy from author's collection. Background: 1978 Tuesday Taylor Penthouse by Ideal, $150.

Doll: 1962 Brunet Ken®, $50. Clothes: 1961 Saturday Night Date, $50; 1964 Country Clubbin', $70; and 1964 Fraternity Meeting, $50. Scene: 1964 Wardrobe by Suzy Goose, $75.

I've just shaved and showered because I've got a date. I wonder what I should wear? So far, I have the gray coat, the checkered coat, and the houndstooth sweater.

Doll: 1961
Brunette Ponytail.
Clothes: Mainly
1961 Singing in
the Shower, $35;
1962 Pak clothes,
$10. Scene: 1962
Dream House by
Mattel, $150.

I have to pick the prettiest dress because I have a date with the handsomest guy in school. He's going to take me to the best restaurant in town and then to a movie. It'll be so cool!

Dolls: 1961 Brunette Ponytail in 1961 Sheath Sensation, $50; 1963 Blond Ken® in Pak clothes, $15; 1962 Brunet Ken® in 1961 Yachtsman, $50. Scene: 1962 Dream House by Mattel, $150.

Doll: 1963 Fashion Queen in 1965 Modern Art, $150. Scene: Mostly 1964 New Dream House by Mattel, $350.

Doll: 1963
Fashion Queen in
original outfit.
Scene: 1964 New
Dream House by
Mattel, $350

Now where did I put that purse? There it is! It doesn't match my boots or raincoat, though. I wonder if it goes with anything in my dress. Oh yes, there's some blue in my dress, so I'm safe!

Doll: 1963 Red Bubble Cut. Clothes: 1961 Raincoat, $35; 1967 Color Magic dress, $20, and 1965 Pak purse, $10. Scene: 1964 New Dream House by Mattel, $350

Doll: 1963 Fashion Queen in 1961 Friday Night Date, $75. Scene: 1962 Dream House wall by Mattel, $150, and 1964 Go-Together furniture by Mattel, $50 all.

21

I have just made us a delicious, romantic dinner. You can lick the spoon while I set the table.

Dolls: 1964 Allan® in 1964 Holiday, $150; 1963 Brunette Midge® in 1959 Barbie-Q, $60. Scene: 1964 New Dream House by Mattel, $350.

Dolls: 1963 Red Midge® in 1965 Lunch Date, $40, and 1964 Blonde Swirl Ponytail in 1962 Suburban Shopper, $75. Scene: 1964 New Dream House by Mattel, $350

We just swoon over the hi-fi music of Nat King Cole. I am teaching her the latest dance steps like the stroll and the mashed potato so we can go to the sock hop on Friday night. It'll be the ginchiest!

23

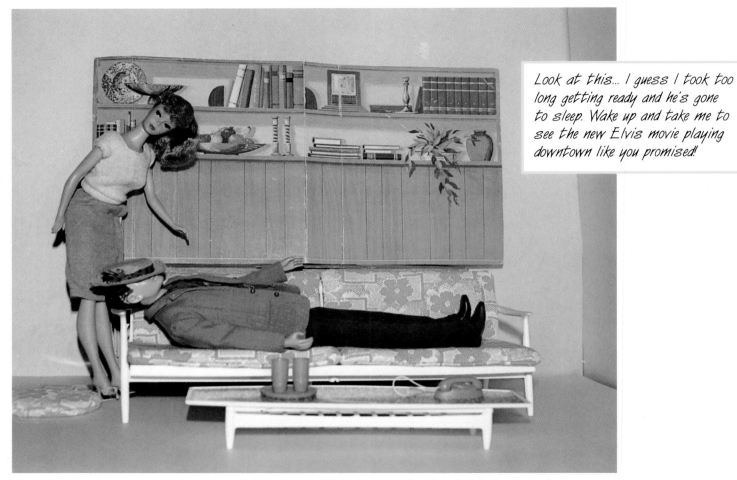

Dolls: 1961 Red Ponytail in 1962 Pak clothes, $10 each, and 1961 Brunet Ken® in 1961 Dreamboat, $60. Scene: 1964 Go-Together Furniture by Mattel, $50.

I didn't know you were coming to the park today, too. I see you have your littlest sister and her dog with you. Isn't the weather nice today? I just love your new outfit; do you like mine? Well, I'll see you tomorrow. Goodbye.

Dolls: 1960 #3 Brunette Ponytail in 1959 Roman Holiday, $1,500; Dog (left) from 1964 Skipper® fashion Dog Show, $175; 1960 Blonde #3 Ponytail in 1959 Gay Parisienne, $1,200; and Brunette Tutti® in 1966 Me & My Dog, $225. Painting by L. Barker; trees from author's collection.

Dolls: 1965 Blonde Bend Leg Skipper® in 1966 Dog Show, $75 (clothes only); Dog from Dog N Duds, $125; and 1965 Blonde Bend Leg in 1965 Poodle Parade, $450. Scene: 1964 New Dream House by Mattel, $350.

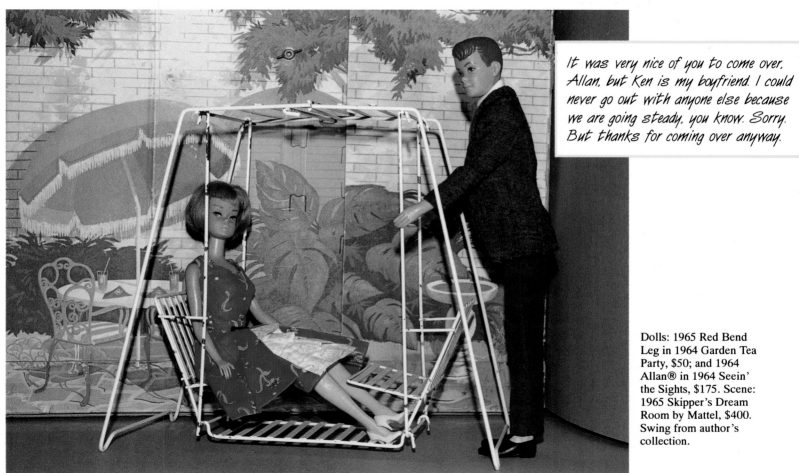

It was very nice of you to come over,
Allan, but Ken is my boyfriend. I could
never go out with anyone else because
we are going steady, you know. Sorry.
But thanks for coming over anyway.

Dolls: 1965 Red Bend
Leg in 1964 Garden Tea
Party, $50; and 1964
Allan® in 1964 Seein'
the Sights, $175. Scene:
1965 Skipper's Dream
Room by Mattel, $400.
Swing from author's
collection.

Look, he wants to go with you. Do you think you could teach him to skateboard?

Dolls: 1965 Red Bend Leg Skipper® in 1964 School Days, $50; and
1964 Red Ponytail Swirl in pink version of 1964 Knitting Pretty, $250.
Scene: 1965 Barbie & Skipper Deluxe House case (Sears DSS), $150

Dolls: 1965 Blonde Bend Leg Skipper® in 1969 Drizzle Sizzle, $75; and 1967 Blonde Twist N Turn in 1967 Fashion Shiner, $100. Scene: 1965 Barbie & Skipper Deluxe House case (Sears DSS), $150.

I am going out shopping for a little while and both of you are staying here with the babysitter. Don't watch too much TV. It's a beautiful day; why don't you go outside and play? If you are very good, I will bring you a surprise!

Dolls: 1965 blonde Bend Leg in 1966 Lunch on the Terrace, $125; 1964 Brunette Skipper® in 1966 Tea Party, $80; 1966 Brunette Tutti® in 1966 Ship Shape, $50; and 1966 Blonde Francie® in 1966 It's a Date, $60. Scene: Barbie Family Deluxe House case, 1966 (Mattel), $150.

I really like your new pantsuit. You can wear it together, or the top can be a dress. Where would you like to go tonight to show off our new outfits? How about a Mod club where they play groovy music like the Beatles?

Dolls: 1970 Red Living Barbie® in 1967 Lemon Kick, $75; and 1969 Brunette Twist N Turn in mostly 1967 Yellow Mellow, $85.
Scene: 1967 World of Barbie House case by Mattel, $100.

31

Your suede pantsuit and polyester shirt look far out! You are really new and good-looking now!

Dolls: 1969 Red Talking in 1968 Fancy Dancy, $100; and 1970 New Good Looking Ken® in 1972 Suede Scene, $75. Scene: 1967 World of Barbie House case by Mattel, $100.

I just love to fill up the tub and put in lots and lots of bubbles. I can soak for days!

Doll: 1971 Blonde Live Action. Scene: 1976 Beauty Bath by Mattel, $35.

Chapter Two

Careers and Conquests: From Actress to Teacher

Our imaginative houses may have been similar, but our interests were very different. Playing with dolls allowed us the luxury of trying out careers and activities as varied as we were, and the toy manufacturers were happy to provide us with outfits and settings for our varied fantasy lives. The usual careers were included in the Mattel line, with the interesting addition of astronaut in 1965, long before NASA ever accepted a woman in its space program (see Chapter 4, History Lessons, p. 86.

Mattel created a Fashion Shop, Little Theatre, and College in the early 1960s. All three are made of cardboard, so they are fragile and rare. These were followed by cafes, boutiques, and beauty salons of vinyl and plastic. Other companies such as Ideal made horse barns and a penthouse with fabulous views—a sure sign of any girl's success if she lived there, and we all knew we would.

We were artists and secretaries, played tennis and skied, took care of the young and the sick, and of course we were teachers. What exciting lives we had back then. Did all of these dream activities give us the confidence to do these things as grownups, or did they reflect our changing times? We are women; hear us roar!

Doll: 1959 #1 Blonde Ponytail. Clothes: 1962 Pak dress, $25; the hat and shopping bag are 1990 Dallas Convention souvenirs. Scene: 1962 Fashion Shop by Mattel, $300.

What I like to do best is shop, of course, but I have many other careers.

As Cinderella, I am stuck in this ugly garage apartment because I am a poor, lonely orphan. My mean old stepsisters make me do all of the housework. I sometimes play the mandolin and make up country-western songs about a handsome prince riding up on his horse to take me away from all this.

36

Doll: 1966 Color Magic, $450. Clothes: 1964 Cinderella (poor), $75. Scene: 1964 Little Theatre by Mattel, $400. Mandolin from author's collection.

And sure enough, he did come by one day. We went to the ball at his palace. Only my tiny and beautiful foot fit into the slipper and he asked me to marry him, which of course I did, so now I'm the princess. Guess what happened to my mean old step-sisters?

Dolls: 1964 Allan® in 1964 Prince, $250; and 1963 Red Midge® in 1964 Cinderella (rich), $150. Scene: 1964 Little Theatre by Mattel, $400.

37

Will you please stop scaring me like this and just tell me where my grandmother is? I have to give her these rolls before they get cold or I will get punished.

Dolls: 1966 Color Magic, $450, in 1964 Red Riding Hood, $250; and Ken®. Mask is part of previous outfit; nightgown was made by author. Scene: 1964 Little Theatre by Mattel, $400.

The best part about being an actress is taking a bow at the end. Here's the whole cast of Little Theatre players: Guinivere, Arthur, Prince, Cinderella (rich and poor), and the Arabian Nights characters.

Various dolls. All costumes from 1964, $100 to $250 each. Scene: 1964 Little Theatre by Mattel, $400. Camel from author's collection.

39

Dolls: 1962 Blond Ken® in 1964 American Airlines Captain, $200; and 1961 Blonde Ponytail in 1961 American Airlines Stewardess, $75. Scene: 1978 Reservation Center (Sears), $50.

Dolls: 1965 Blond Bend Leg Ken® in 1962 Saturday Night Date, $50; and 1965 Blonde Bend Leg in 1966 Pan American Airlines Stewardess, $1,500. Scene: 1958 wooden furniture by Mattel, $75. Background: 1978 Tuesday Taylor Penthouse by Ideal, $150. Luggage from author's collection.

Doll: 1964 Blonde Swirl Ponytail in 1965 Pak dress, $75; and
1963 Baby from Barbie Babysits, $150. Scene: 1963 Kitchen
(Deluxe Reading), $175. Accessories from author's collection.

Don't be scared, honey. Just point your toes and they'll love you!

Dolls: 1962 Red Ponytail in 1961 Ballerina, $50; and 1964 Red Skipper® in 1964 Ballet Class, $50. Scene: 1977 Sonny & Cher Theatre in the Round by Mego, $250.

43

Your hair is a mess! First, I need to comb it, then wash it, then roll it up so you can sit under the dryer for an hour. After that, I will tease it up real high and spray it, so it should last for the whole week!

Dolls: 1961 Blonde Bubble Cut in 1961 Registered Nurse, $50; and 1964, Blonde Ponytail Swirl in 1966 Color Magic separates, $25. Scene: 1976 Beauty Salon (Sears), $50.

44

Dolls: Various Bubble Cuts, 1961-65. Clothes: Various Pak items from 1962, $10 each. Scene: 1964 College Campus (Sears DSS), $400.

Yes, I do like you and yes, I did mean it when I offered to carry your books after class, but...

Dolls: 1964 Brunette Bubble Cut in 1966 Beau Time, $150; and 1962 Brunet Ken® in 1965 Going Bowling, $50. Scene: 1964 College Campus (Sears DSS), $400.

It's great to take a break from studying and come down to the Malt Shop. What did you say your job is, jerk? Oh, Soda Jerk!

Dolls: 1965 Bend Leg Allan® in 1964 Fountain Boy, $75; and 1965 Red Midge® in 1967 Tropicana, $50. Scene: 1964 College Campus (Sears DSS), $400.

47

It's so romantic here at the Malt Shop. We can hold hands, sip our Cokes, and listen to the Supremes on the jukebox.

Dolls: 1965 Blonde Bend Leg in 1965 Fun N Games, $100; and 1965 Bend Leg Ken® in 1962 Pak separates, $5 each. Scene: 1964 College Campus (Sears DSS), $400.

A toast! It's so exciting working in the big city, having our own apartment, and wearing the latest fashions! Now we are even luckier because one of us just got a raise! Can you guess which one?

All Bend Leg dolls, fashions from 1965: (left to right) Red Midge® in Saturday Matinee, $500; Brunette Barbie® in Matinee Fashion, $350; Blonde Barbie® in Shimmering Magic, $850; Blonde Barbie® in On the Avenue, $350; Blonde Barbie® in Music Center Matinee, $450; and Red Barbie® in Gold N Glamour, $750. 1958 Furniture by Mattel, $100 for all. Background: 1978 Tuesday Taylor Penthouse by Ideal, $150. Accessories from author's collection.

Doll: 1961 Blonde
Ponytail in 1961
Busy Gal, $225.
Various separates.
Scene: 1962
Fashion Shop by
Mattel, $350.

Here are some new
designs I've drawn.
Would you like to
see them?

I think this dropped waist dress with the big bow is adorable. I'll take this one.

Doll: 1964 Brunette Bubble Cut in 1963 Career Girl, $150, holding 1965 Pak dress, $75. Various separates. Scene: 1962 Fashion Shop by Mattel, $350. Accessories from author's collection.

Notice my new dress, gloves, and matching straw hat and bag. They are just like the latest Paris fashions I saw in the magazines. My outfits are always coordinated, an outfit for every activity. Today I am not working as a Fashion Model, so I am going shopping! I just love new clothes, don't you?

Doll: 1960 #3 Blonde Ponytail in 1959 Plantation Belle, $250. Scene: 1962 Fashion Shop by Mattel, $350.

52

He's a great horse. His name is Dancer and he can slow dance or jitterbug across the meadows. Are you ready for your riding lesson?

Dolls: 1963 Brunette Bubble Cut in 1966 Riding in the Park, $375; and 1965 Brunette Skooter® in 1966 Learning to Ride, $175. Horse: 1971 Dancer by Mattel, $175. Scene: 1975 Jody's Horse Farm by Ideal, $75.

53

Dolls: 1962 Brunette Ponytail in 1962 Ice Breaker, $50; 1962 Brunet Ken® in 1962 Fun on Ice, $50; and 1964 Brunette Skipper® in 1964 Skating Fun, $65. Scene: 1964 Little Theatre by Mattel, $400.

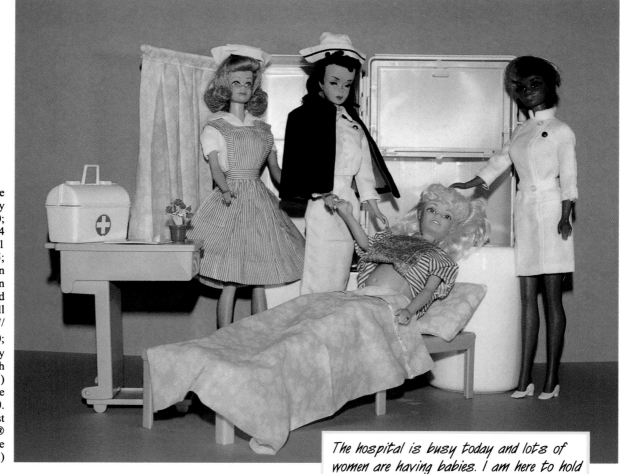

Dolls: 1963 Blonde Midge® in 1964 Candy Striper Volunteer, $250; 1960 Brunette #4 Ponytail in 1961 Registered Nurse, $75; 1970 Julia (Diahann Carroll) by Mattel in original outfit, $125; and Mommy To Be Doll (Not by Mattel Reg #07/883,500 ©1988, $50; This doll has a tummy compartment which houses a tiny baby!) Scene: 1988 Loving Care Playcase by Mattel, $50. (This was the first hospital, but Barbie® became a doctor in the mid-1970s.)

The hospital is busy today and lots of women are having babies. I am here to hold your hand, so don't worry about a thing.

55

Today's lesson will be geography. Do you know where Scotland is? Please stop pulling her hair and pay attention or you will not know the answer for the quiz!

Dolls (left to right): 1965 Ricky® in Saturday Show, $50; 1964 Blonde Skipper® in What's New at the Zoo, $65; Red #5 Ponytail Barbie® in 1965 Student Teacher, $175; 1965 Brunette Skooter® in Rainy Day Checkers, $175; and 1964 Brunette Skipper® in School Girl, $150. Scene: 1996 Classroom Furniture by Arco, $15; and 1996 Blackboard by Mattel, $50 with three-doll set. Background: 1965 Skipper Dream Room by Mattel, $400.

Life at the top can be so wonderful, but it's not carefree. For instance, I still have to watch the cat so she doesn't eat the fish. Also, I am so high up, who is going to clean my windows?

Doll: 1965 Red Bend Leg in 1967 Intrigue, $125. Scene: 1958 furniture by Mattel, $50 all. Pets are Avon bottle, circa 1980s. Background is 1978 Tuesday Taylor Penthouse by Ideal, $150; 1964 rug from New Dream House by Mattel, $350

57

Her first car was an orange Austin Healy with aqua interior, circa 1962, made by Irwin Corp. I have not been able to stop her since. Later models made by Mattel, Inc. include a dune buggy, various Corvettes, a 1957 Chevy, Jeep, Jaguar, Ferrari, and a Mustang. She has had airplanes (both single engine and commercial), many boats and yachts, campers, buses, motorhomes, bikes, and cycles. The list is much shorter for what she has *not* had: a train, space capsule, helicopter, or covered wagon. All of these were made by other companies such as Kenner, Hasbro, and Ideal and are really neat accessories!

Wheels give us freedom, and Barbie® doll goes to all the fun places. There were costumes from many countries in the 1960s, then a beautiful line of International Dolls of the World Series was begun by Mattel in 1980 as department store specials (DSS) and continues today. Every country imaginable is represented.

It's probably not an accident that the preferred method of storage for our dolls was in a carrying case. We were on the go, and the whole world was out there waiting for us!

On the first weekend at college, I drove my nifty sports car to the football game. I met the cutest player and he asked me out for next Friday night!

Dolls: 1962 Fashion Queen in 1959 Winter Holiday, $75; and 1961 Ken® in 1961 Touchdown, $60. Car: 1962, by Irwin, $175. Scene: 1964 College Campus (Sears DSS), $400.

So we went to the drive-in movie on Friday night. A really cute guy parked right next to us and flirted with me. I couldn't keep my eyes off him. My date said just to ignore him, that he was probably in a gang because of his jeans and huge, loud motorcycle.

Dolls: Two 1965 brunet Bend Leg Ken® dolls — on left wearing 1964 Going Hunting, $50, on right in 1965 Bend Leg shirt, $50; and 1964 Blonde Bubble Cut in 1962 Pak shirt, $10. Cycle: 1993, by Toy State, $30. Car: 1962, by Irwin, $175. Scene: 1964 College Campus (Sears DSS), $400.

Of course I didn't listen. I snuck out of the dorm after curfew and met him at the drive-in. He turned out to be a swell guy, not a delinquent after all, and we have a date for next Friday night. I will get to ride on the back of his motorcycle!

Dolls: 1964 Blonde Bubble Cut in 1962 Pak separates, $10 each; and 1965 Brunet Bend Leg Ken® in 1964 Going Hunting, $50. Cycle, 1993, by Toy State, $30.

Dolls: 1961 Blond Ken® in 1964 Ken A Go Go, $450; and 1964 Blonde Ponytail in 1962 Pak shortsuit, $10. Car: 1963, by Irwin, $175. Scene: World of Wonder Class Act Folder.

I just love college life. I met another guy. I mean really handsome, and he has a hotrod with real loud side pipes and glass packs in his muffler. I can hear him all the way in the back of the dorm when he comes to pick me up. We cruise down the drag looking for other couples in cool cars. Whee!

Dolls: 1964 Blonde Ponytail in 1962 Pak dress, $20; and 1963 Brunette Midge® in 1964 Outdoor Art Show dress, $250. Pak purses, 1962, $10 each. Scene: Paloma perfume store display.

I just love Modern Art, don't you? I wonder what it is?

Dolls (left to right): 1965 Brunette Skooter® in 1964 Land & Sea, $80; 1965 Blonde Skipper in 1964 Ship Ahoy, $110; 1964 Blonde Bubble Cut in 1964 Aboard Ship, $175; and 1965 Brunette Bend Leg Midge® in 1959 Resort Set, $60. Scene: 1975 Dream Boat case by Mattel, $150.

Doll: 1982 Eskimo DSS, $150. Painting by Jean
Barney. Bear and accessories from author's collection.

I rode my camel across the desert just to see you. You are the most beautiful woman I have ever seen in your pink scarf outfit. Come away with me to my nomad village where my tent is adorned with jewels and your every wish is my command!

Dolls: 1964 Brunette Ponytail in 1964 Arabian Nights, $200; and 1962 Brunet Ken® in 1964 Arabian Nights, $100. Scene: 1964 Little Theatre by Mattel, $400. Camel from author's collection.

Here I am on the French Riviera, or the Cote d'Azure as they say. The beaches are pebbles and the water is cold, but the French men are fantastic! Alors!

Doll: 1965 Brunette Bend Leg Midge®. Purse from Suburban Shopper, 1959, $25. Scene and accessories from author's collection.

We are can-can dancers at a club in Montmartre. We have the Eiffel Tower, beautiful chateaus, and lots of fun parties because the night life in Paris is formidable!

Dolls: 1997 French, $25; and 1980 Parisian DSS, $175. Scene and accessories from author's collection.

Here, come and taste this pineapple I have picked for you. It's so sweet and delicious. Isn't Hawaii a real paradise?

Dolls: 1964 Brunette Swirl Ponytail in 1964 Hawaii, $100; and 1962 Ken® in 1964 Hawaii, $75. Palm trees from Hammock Hideaway, 1991, by Mattel, $40. Painting by Jim Morrison. Accessories from author's collection.

Dolls: 1965 Midge® , Wig Wardrobe, $200, in 1964
Holland, $125; and 1964 Allan® in 1964 Holland,
$100. Accessories from author's collection.

The Orient is so exciting, especially Japan where I was first made. The customs are mysterious but fascinating. The music is so different. They have gorgeous temples, and they use lots of red. I think it stands for riches and good luck!

Dolls: 1985 Japanese DSS, $150; and 1962 Fashion Queen with brunette wig in 1964 Japan, $250. Accessories from author's collection.

We're in colorful, sunny Mexico. He's just a tiny little bull, and he's so adorable. How can you say you want to fight with him?

Dolls: 1963 Brunette Bubble Cut in 1964 Mexico, $125; and 1962 Ken® in 1964 Mexico, $125. Scene: 1964 Little Theatre by Mattel, $400. Bull, 1980, by Fisher-Price, $5.

Dolls: 1981 Scottish DSS, $150. Piper is a souvenir from International Gifts, circa 1994. Scene: 1964 Little Theatre by Mattel, $400. Chair from author's collection.

73

We are carefree gypsies from Spain. We sing and dance around our colorful campsite. Cross my palm with silver and I will tell your fortune (just for fun, of course)!

Dolls: 1980 Hispanic Barbie®, $150; and 1983 Spanish DSS, $150.
Scene: Esmerelda Gypsy Festival Tent, 1995 Hunchback/Notre Dame series by Mattel, $25. Pink cart from author's collection.

74

Dolls: 1961 Ken® in 1964 Switzerland, $100; and Blonde #4
Ponytail in 1964 Switzerland, $150. Scene: 1964 Little Theatre
by Mattel, $400. Furniture from author's collection.

75

Dolls (left to right): 1962 Ken® in 1962 Ski Champion, $65; 1961 Brunette Ponytail, in 1963 Ski Queen, $75; 1974 Sun Valley Barbie®, $50; and 1976 Gold Medal Skier Barbie® and Ken®, $50 each. Scene: 1972 Mt. Ski Cabin (Sears DSS), $75.

Chapter Four

Ancient Greece. Arthur and Guinevere's England. Scarlett and Rhett in Atlanta. The West. These visions have always beckoned me. When I saw a movie about them, I would re-enact it in my mind for days, imagining what life was like in another time. Then I would recreate it with my dolls. Were these my past lives? No wonder I made perfect grades in history class. No wonder that now my ideal vacation is to go to these places and see history firsthand, these countries and times I've read about, studied about, dreamed about all my life. To walk where Socrates walked, stand under a tree where history was made, see the same vistas as the great ... These things still send shivers up my spine!

On a beautiful day like today I wander the hills of Greece. With an ancient temple in the background, I find a column and play my bouzouki. Then I stretch out for a nap and dream of the gods and goddesses who lived on Mt. Olympus. Were they real?

Dolls: 1986 Greek DSS, $100; and Great Eras Series, 1996 Grecian Goddess, $100. Accessories from author's collection.

I am your devoted Guinivere, King Arthur. I love you with all my heart and I do not want you to go off to war in these Middle Ages. But you must do what you think is right. Ask the knights of your Round Table what they think. Why not take a vote?

Dolls: 1963 Blonde Bubble Cut in 1964 Guinivere, $175; and 1961 Ken® in 1964 King Arthur, $175. Scene: 1964 Little Theatre by Mattel, $400.

79

Anna and the King of Siam, now Thailand, whirl around and around in 1860 as they practice the waltz Soon there will be dignitaries coming by sailing ship. The King has many, many children, and Anna came from Britain to be their teacher. See the littlest one peeking through the plant to watch?

Dolls: 1961 Ken® in 1964 Arabian Nights, $100; and 1961 Redhead Ponytail in original outfit made by author. Plant, 1964 by Mattel, $10. Scene: 1964 Little Theatre by Mattel, $400. Tiny Thai from author's collection.

My name is La Goulou. It is 1898, and I work at the Moulin Rouge in Paris. The famous artist, Toulouse Lautrec, comes here often to sketch me as I dance. Here we invented the can-can where I lift my skirt and kick my leg so high that you can see my bloomers. How naughty we are!

Doll: 1960 Blonde #4 Ponytail in original outfit made by author. Accessories from author's collection.

The lake is so still at sunset. I feel the warm breezes turning cooler as I glide over the water in my birch bark canoe. It's the right time to look for little deer as they come out to hunt for food. What a perfect setting in our land which the English settlers call America.

Doll: 1983 Phoenix Convention Doll made by Ellen Altig, $200. Bow and sling by Bob Young, $25. Painting by Genie Askins. Accessories from author's collection.

Dolls: 1995 Pilgrim Barbie®, $25, and 1982 Michigan Convention Doll designed by Carol Spencer for Mattel, $350. Painting unsigned. Accessories from author's collection.

83

We are pioneer women, moving across the Great Plains in our wagon train along the Oregon Trail. I think I would rather go to California, though. I hear they found gold in the hills and lots of people will get rich. Besides, they're going to invent movies soon and then I could be a star!

Dolls: 1983 Phoenix Convention Doll designed by Janet Goldblatt for Mattel, $350; and 1996 Little Debbie mail-in doll, $50. Horse, wagon, and scene, 1995 by KidKore, $30.

84

Dolls: 1960 Brunette #3 Ponytail in original outfit made by author; and 1983
Angel Face Barbie®, $50. Scene: Jody's Victorian Parlor, 1975, by Ideal, $75.

It's a great day for the U.S. in 1969. We have finally landed on the moon. Where should I put the flag? Do you think it would look better here or over there?

Dolls: 1962 Ken® and 1960 #4 Ponytail, both in 1965 Astronaut, $350 each. Six Million Dollar Man Spaceship by Kenner, 1975, $125.

86

Dolls: 1990 Gorby by Dreamworks, $75; and 1989 Russian DSS, $125. Scene of Corn Palace in Mitchell, South Dakota. Turret from author's collection.

I can see why they've called it the Cold War. Russia is freezing! You really are a tough little man, Gorbachev.

Chapter Five

The year is packed with holidays! Each month has at least one, and some even have three or four! I love to dress up my dolls for these days and remember the special times we are celebrating. To begin, there's New Year's, Valentines, and St. Patrick's, plus a few some people might not even know about such as Mardi Gras and Go Western Days. So here we go. Just like the character Truvy in *Steel Magnolias*, we decorate for everything!

Dolls (left to right): 1969 Brunette TNT in 1968 Extravaganza, $175; 1969 Talking Ken® in 1969 Guruvy Formal, $75; 1961 Ken® in 1961 Tuxedo, $75; 1960 Brunette #4 Ponytail in 1960 Enchanted Evening, $175; 1977 Superstar Ken® and Barbie® in original outfits, $75 each; and 1984 Crystal Barbie® and Ken® (Don't they look young on their 25th anniversary!), $50 each. Scene: 1985 Dinner Date by Arco, $75.

We're swinging and swaying in the New Year with all of our friends!

Dolls: 1964 Red Swirl Ponytail in 1965 Junior Prom, $300, and 1962 Blond Ken® in 1964 Best Man, $750. Scene: 1985 Dinner Date by Arco, $75.

Dolls: 1990 Dallas Convention doll by Mattel, $300; and 1967 Brunette TNT in mostly 1971 Gaucho Gear, $100. Scene: 1975 Jody's General Store by Ideal, $75.

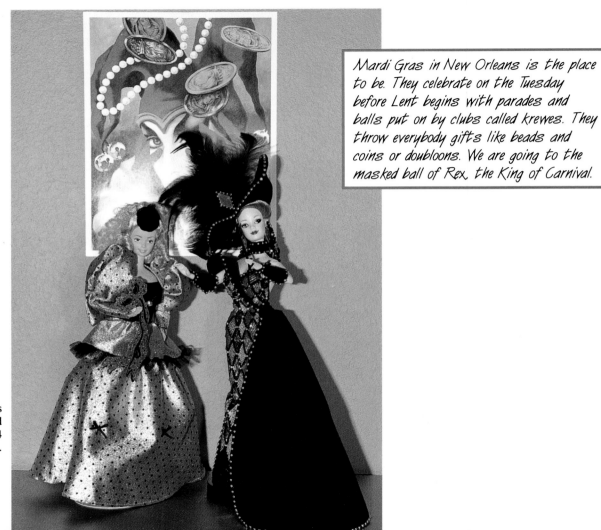

Dolls: 1988 Mardi Gras
Barbie® DSS, $75; and
Bob Mackie's 1994
Masquerade, $450.

Mardi Gras in New Orleans is the place to be. They celebrate on the Tuesday before Lent begins with parades and balls put on by clubs called krewes. They throw everybody gifts like beads and coins or doubloons. We are going to the masked ball of Rex, the King of Carnival.

Come out, little men. It's Saint Patrick's Day!

Dolls: 1984 Irish DSS, $125; and 1995 Irish reissue, $30. Figures from Burger King's 1994 Snow White series, $5 each. Scene from author's collection.

Dolls: 1972 Walk Lively Barbie®, $125, in late 1980s Toys R Us costume, $20; and 1994
Blonde doll in costume from World Doll Costumes, $10. Accessories from author's collection.

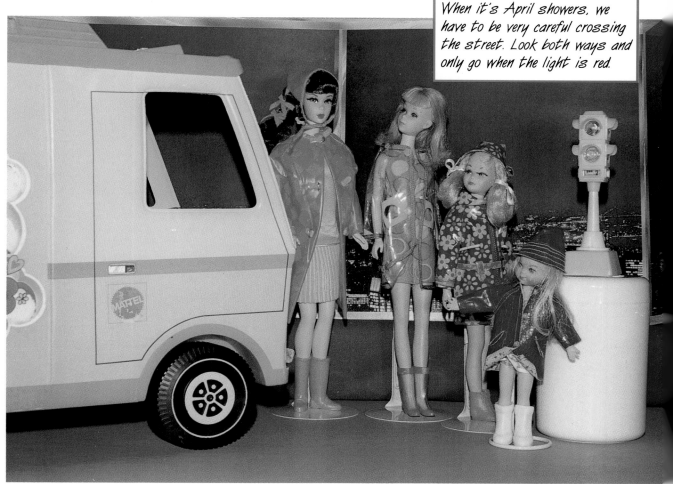

When it's April showers, we have to be very careful crossing the street. Look both ways and only go when the light is red.

Dolls: 1968 Brunette Talking, $150, in 1968 Drizzle Dash, $50, and 1968, Disco Dater dress, $125; 1966 Blonde Francie®, $100, in 1968 Pazam, $125; 1970 Blonde Living Skipper®, $30, in 1967 Flower Showers, $45; and 1966 Blonde Tutti in 1966 Puddle Jumpers, $30. Camper, 1977, by Mattel, $30. Background: 1985 Dinner Date by Arco, $75. Accessories from author's collection.

Dolls: 1964 Black Bubble Cut in 1963 Graduation, $50; 1964 Blonde Skipper®, in 1964 Red Sensation, $40; and 1964 Blonde Swirl Ponytail in 1965 Pak dress, $75. Scene: 1964 College Campus, (Sears DSS), $400.

Graduation Day is finally here in May. I have worked so hard and now I have the diploma with my degree. I am ready for the real world!

We went to the prom and danced until we couldn't stand up anymore. Afterward we invited the guys up for coffee and I am getting a foot rub, but it's time for them to leave. My girlfriend and I share this apartment and we have to get up early to go to our new jobs tomorrow. I hope they will call us again and ask us out on another fabulous date, which they will pay for again, of course!

Dolls: 1964 Red Swirl Ponytail, in Pink Sparkle, $125; 1962 Ken® in 1961 Tuxedo, $75; Ken in 1964 Best Man, $750; and 1965 Brunette Bend Leg in 1967 Evening Enchantment, $350. Scene: 1964 New Dream House by Mattel, $350. Accessories from author's collection.

June is the perfect month for brides. The weather is warm, the flowers are in bloom, and school is out, so all of the little ones can be in my dream wedding!

Dolls: 1960 Blonde #4 Ponytail in 1959 Wedding Day, $150; 1965 Ricky® in Sunday Suit, $50; 1962 Ken® in 1961 Tuxedo, $75; 1964 Skipper® in 1966 Junior Bridesmaid, $200; and 1966 Tutti® in 1966 Flower Girl, $125. Scene: 1964 Little Theatre by Mattel, $400.

Doll: 1964 Brunette Bubble Cut in 1966 Here Comes the Bride, $300. Scene: 1964 New Dream House by Mattel, $350. Luggage from author's collection.

In July, I love to go fishing. I wear my dungarees and my big straw hat so I won't get any freckles. Won't everybody be surprised when they see this whopper I caught!

Doll: 1964 Brunette Ponytail in 1959 Picnic, $125. Scene: 1964 Little Theatre by Mattel, $400. Boat from author's collection.

July 4th is for picnics. We've found this perfect spot out in the country beside a little pond with swans and ducks. It even has a swing hanging from this big old green tree. Whee!

Dolls: 1964 Brunette Skipper® in 1966 Country Picnic, $250; 1961 Ken® in Sport Shorts, $25; and 1960 Blonde #4 Ponytail in 1964 Lunch Date, $35. Scene: Tree from 1975 Young Sweethearts Wishing Well Park by Mattel, $150. Painting by Myrtle Barrett.

August is when the baseball playoffs happen. I just love to go and watch the men play. My boyfriend is the pitcher and the best batter on the team. He hits a homerun at almost every game. We'll eat hot dogs afterward and decide where to go on our date tonight.

Dolls: 1964 Red Skipper® in 1966 Can You Play, $50; 1964 Red Ponytail in 1962 Pak separates, $25 each; 1962 Brunette Bubble Cut in 1962 Pak separates, $10 each; 1965 Ricky® in 1965 Little Leaguer, $50; and 1962 Ken® in 1962 Play Ball, $50. Hot Dog Stand, 1988, by Mattel, $50. Plants from author's collection.

Sitting on the patio and sunbathing is really my favorite sport in August. Look, here come my next door neighbors. I'll bet they want me to barbecue some hamburgers. They are triple the fun!

Dolls: 1960 Blonde #4 Ponytail in original swimsuit; and three hair shades of Midge® doll, 1963, in original swimsuits. Scene: 1964 New Dream House by Mattel, $350. Plant from author's collection.

A tall pink lemonade and then a snooze in the lawn swing while I listen to my favorite deejay ... my idea of the very best summer afternoon before a date!

Doll: 1964 Miss Barbie® in original outfit, swing, and accessories $400. Plants from author's collection.

Come on, it's beautiful outside in the backyard. Stop watching that silly soap opera on TV and go swimming with me! It's our last week of fun before school starts again.

Dolls: 1965 Brunette Bend Leg Midge®, in Pak suit, $10; and 1965 Bend Leg Allan® in original suit, $20. Scene: 1965 Barbie and Skipper Deluxe Dream House (Sears DSS), $150.

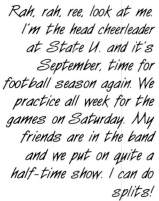

Rah, rah, ree, look at me. I'm the head cheerleader at State U. and it's September, time for football season again. We practice all week for the games on Saturday. My friends are in the band and we put on quite a half-time show. I can do splits!

Dolls: 1963 Red Midge® in 1964 Drum Majorette, $75; 1963 Brunette Bubble Cut in 1964 Cheerleader, $75; and 1964 Allan® in 1964 Drum Major, $75. Scene: College Campus, 1964 (Sears DSS), $400.

Dolls: 1964 Brunette Skipper® in Masquerade, $50; 1966 Blonde Tutti® in Clowning Around, $75; 1960 Brunette #4 Ponytail in 1963 Masquerade, $75; and 1964 Ken® in Masquerade, $75. Scene: 1964 Little Theatre by Mattel, $400. Feed Me™ is a Milton Bradley game.

107

On Thanksgiving Day, we cook a turkey and my favorite vegetables. Then we have our boyfriends over. They are in the service and have just come home on leave. Everything is beautiful and perfect except that they will be leaving again tomorrow. Boo hoo!

Dolls: 1964 Allan® in Sailor, $50; 1963 Blonde Midge® in 1960 Let's Dance, $75; 1962 Brunette Ponytail in After Five, $50; and 1962 Brunet Ken® in Army/Air Force, $75. Late 1980s wood furniture (Sears DSS), $100 all. Table and chairs for Sindy doll by Marx, late 1980s, $25. Background: 1978 Tuesday Taylor Penthouse by Ideal, $150

Decorating for the Winter Carnival is so much fun. We build a giant snowman and watch everyone slow down and grin as they pass. I wish he'd come to life just like in the song!

Dolls: 1969 Brunette TNT in 1972 Poncho Put-ons, $50; early 1980s Snowman by Ideal, $50; 1963 Brunette Skipper® in Sledding Fun, $150. Sled, bench, and trees from author's collection.

The holidays are coming and my yard is entered in the outdoor decorating contest. I have a big red sled full of toys, but no Santa. Maybe he's delivering our presents in the house! I believe, honest I do!

Doll: 1982 Pink & Pretty Barbie®, outfit #4277, $50. Sled from 1987
Oklahoma Convention, $25. Accessories from author's collection.

Dolls: Santa Ken® and PJ-clad Barbie®, gifts from 1987 Oklahoma Convention, $100 each. Sofa made by Mary Eads. Other accessories from author's collection.

On Christmas morning we wake up to a room full of presents under the tree! We each got just what we wanted and a whole lot more because we were very, very good. Well, most of the time!

Dolls: 1964 Red Skipper® in 1971 Dressed in Velvet, $25; and 1971 Live Action Barbie in 1971 Cloud Nine, $40. Accessories from author's collection.

Chapter Six

She's so popular, just like we wanted to be! Since 1961, her boyfriend has been Ken® doll. In 1963, her best friend was named Midge, but this expanded to include Stacey, Casey, Jamie, Christie.... Then there were sister Skipper and her friends, Skooter, Ricky, Fluff, Tiff, Ginger, cousin Francie, and Lori and Rori, Angie and Tangie, and Nan and Fran. The list goes on and could be a Jeopardy game show category. Twin siblings Tutti and Todd® were the tiniest of the bunch, introduced in 1966, and they are so cute!

There was a wonderful time in the 1970s when toy makers created 11.5- to 12-inch dolls for many movie and TV celebrities. Ideal, Kenner, Hasbro, Mego, and others created licensed dolls resembling such stars as Cher (with more than 60 outfits designed by none other than Bob Mackie!), Diana Ross, Farrah Fawcett, Dolly Parton, and the bionic couple, Jamie Somers and Lee Majors (remember them?). Mattel made Twiggy and Diahann Carroll, and Barbie® doll herself aspired to be a superstar. Didn't we all?

I had to have them all, of course, and I'll bet you had a few of them too—or wanted them—because we can never have too many friends!

Exterior of 1965 Skipper Dream Room by Mattel, $400. This is all cardboard and very hard to find.

Dolls: 1965 Brunette Skooter® in Lounging Lovelies, $50; and 1965 Blonde Bend Leg Skipper® in Dreamtime, $50. Scene: Interior of Skipper Dream Room.

Dolls: 1967 Blonde Chris®, $90, in Let's Play Barbie, $175; and 1965
Bend Leg Blonde Skipper® in 1965 Me N My Doll, $125. Scene:
Interior of Skipper Dream Room. Case on bed by Rebecca Brosdahl.

Exterior of 1966 Tutti and
Todd House (Sears DSS),
$175. Vinyl-over-cardboard
case. Very hard to find.

Dolls: 1968 Buffy and Mrs. Beasley, $125; 1966 Todd® in Sundae Treat, $100 for dressed doll only. Scene: Interior of Tutti and Todd House. Balloons from author's collection.

Brunette 1966 Tutti® in Ship Shape, $50. Pool, no
date, by PlaySkool, $5. Toys from author's collection.

Francie House, 1966, by Mattel, $125. Vinyl-over-cardboard case. Her first house.

120

Dolls: 1972 No-Bangs Francie® in 1967 Cool White, $100; 1969 Twist N Turn Francie® in 1968 Mini Chex, $50; and 1967 Eyelash Francie® in It's a Date, $50. Scene: Interior of Francie House. Birdcage from author's collection.

I got up the courage to call him on the phone, and guess what? His mother answered. I was scared she wouldn't like girls calling him, so I hung up!

Dolls: Two 1970 Francie® Hair Happenings in 1966 Dance Party, $125, and 1969 The Yellow Bit, $125. Scene: Interior of Francie House.

Talking Brad®, 1971; and The Malibu series dolls: 1973 Christie®, 1972 PJ®, 1971 Skipper®, 1971 Barbie®, 1971 Ken®, and 1971 Francie®. Scene: 1965 Skipper Dream Room by Mattel, $400, and 1984 Bubbling Spa by Mattel, $25.

Exterior of 1971 Barbie Cafe Today by Mattel, $300. Vinyl-over-cardboard case with plastic roof. Very hard to find.

124

I don't believe it. Look who's taking the orders at the drive-in window. It's none other than the Pink Panther!

Dolls: 1984 Pink Panther from United Artists, $25; 1969 Brunette TNT in original suit; and 1969 Talking Ken in original suit. Sun N Fun Buggy, 1971, $75. Scene: 1971 Cafe Today. Shrub from author's collection.

125

This is a really far-out cafe. We come here to drink coffee and listen to poetry written by our friends. Sometimes there is a folk singer playing the guitar. This is the scene, man, and we are cooool, man, real cool.

Dolls (left to right): 1971 Live Action Ken® in 1969 Big Business, $50; 1968 Blonde Talking Stacey® in 1967 Zokko, $100; 1969 Red Stacey® in 1967 Plush Pony, $75; 1968 Blonde TNT Stacey® in 1967 Swirley-Cue, $125; 1971 Live Action Christie® in original outfit, $150; 1970 Brad® in 1969 Town Turtle, $50; and 1970 Christie® in 1967 Flower Wower, $40. Scene: 1971 Cafe Today.

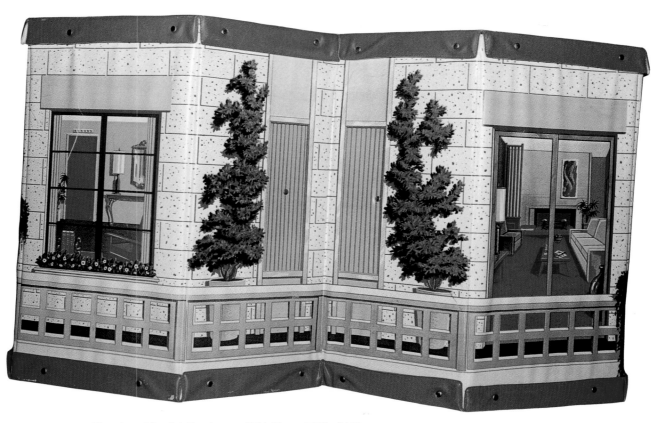

Exterior of Jamie's Penthouse, 1971 (Sears DSS), $400.
Vinyl-over-cardboard case with plastic furniture. Very rare.

It's for the dog!

Dolls: dog from 1965 Dog N Duds, $125; 1972 Walk Lively Barbie® in original suit, $175; and 1972 Walking Jamie®, $150, in 1967 Dreamy Blues, $40. Scene: 1971 Jamie's Penthouse, 1971.

This dinner looks delicious. I'm sure glad you can cook. All I can make is beenie weenies and macaroni out of a box, but I'll invite you over to my place next week. Maybe we can get Chinese food?

Dolls: 1972 Walk Lively Steffie® in original suit, $125; 1972 Walk Lively Ken® in original clothes, $50. Scene: 1971 Jamie's Penthouse. Food from author's collection.

129

I am so excited to be a guest on the Sonny and Cher Show. We are at rehearsal and Cher is going to sing another one of those sexy ballads!

Dolls (left to right): 1977 Farrah by Mego, $50, in Toys R Us outfit, $25; 1971 Red Living Barbie® in 1969 Firelights, $50; 1977 Jaclyn Smith by Mego, $75, in original suit; 1976 Growing Hair Cher by Mego, $75, in 1975 Starlight, $25; and 1976 Sonny by Mego, $50 in original clothes. Scene: 1977 Sonny and Cher Theatre in the Round by Mego, $250. Very rare.

Dolls: 1972 Busy Ken®, $50 in original clothes; 1976 Sonny by Mego, $50 in original clothes; and 1971 Living Barbie® in 1971 Peasant Dressy, $75. Scene: Sonny and Cher Theatre in the Round. Furniture from author's collection.

In this skit, Sonny has a pizza parlor. He trips and spills it all over us and we get mad, but he is so cute with his Italian accent that we wind up laughing hysterically!

Dolls (left to right): 1977 Tenille by Mego, $75; 1977 Diana Ross by Mego, $125; 1960 Blonde #4 Ponytail in Solo in the Spotlight, $150; and 1976 Cher by Mego, $50 in Mint Julep, $35. Scene: Sonny and Cher Theatre in the Round.

Can you believe it? These stars are singing backup harmony for me!

132

Thanks so much for having me on your show. This has been my dream come true! Will you invite me back to guest star again if I learn a new song? Hah, gotcha!

Dolls: 1967 Blonde TNT in 1967 All That Jazz, $125; and 1976 Cher by Mego, $50, in 1976 Montgomery Ward DSS outfit, $25. Scene: Sonny and Cher Theatre in the Round.

Exterior of Barbie and
Skipper Vanity Trunk,
1965 (SPP), $250.
Vinyl over cardboard
with plastic furniture
inside. Very rare.

Dolls: 1960 Blonde Ponytail #4 in 1965 Slumber Party, $150, and 1964 Blonde Skipper® in Dreamtime, $50. Scene: Interior of 1965 Barbie and Skipper Vanity Trunk.

135

Chapter Seven

Pricing

In the past five years, Barbie® doll collecting has boomed, becoming so popular that new dealers and even speculative investors are now attracted to the hobby. She's come a long way from simply being our favorite playpal and is now *THE* hot doll, surpassing even the long-entrenched Mme. Alexanders and Shirley Temples. Prices of everything associated with the dolls continue to rise at an incredible rate with no peaks in sight for the vintage dolls, clothing, and structures. The only good a price guide will do is for comparison, to show which things have more value than others.

Many books quote prices for a never-removed-from-box (NRFB) or mint-in-box (MIB) condition, and these items command exceptionally high prices because they are so rare. These prices give an unrealistic impression of what most items found by collectors can be bought and sold for. The NRFB prices should be halved for an excellent doll without the box.

The prices I suggest are for dolls that have been played with but are in very good to excellent condition, the status of most of my collection. Dolls which are soiled or that have been repaired, touched-up, had their hair redone, or been cleaned in any way are not original. They should be carefully labeled and valued at 20 to 50 percent less. Especially troublesome are dolls with green ears, the result of earrings left in the head for years. These came with the early dolls but should be removed and not put back. Collectors put the earrings in the hair for short periods of time or replace the posts with stainless steel ones. There are several methods of removing the green with chemicals but none are safe. With some, the green returns; with others it ruins the whole head, so these dolls are risky investments. Dolls in poor shape must be sold at garage sale prices no matter how old they are.

Barbie® Herself

Barbie® dolls fall into three categories determined by the predominant face mold used: Vintage or Ponytail (1959-66), Eyelash (1967-76), and Superstar (1977 to present). Many hairstyles were

Three faces of Barbie®
(left to right): Ponytail or
Vintage, 1959-1966;
Eyelash Era, 1967-1972
(1976); and Superstar,
1977-present.

made for each head mold because Mattel, Inc.had to come up with something new every year to offer varirty and sell more dolls. Ponytail and Eyelash have straight arms and closed mouths; Superstar has arms bent at the elbows and a toothy smile. There are exceptions to this, such as the Malibu and My First Barbie lines, but these are a good indicator of age. The year on her hip has *nothing* to do with age because it shows the copyright year of the body and molds were used for years, sometimes decades.

The first face mold was named for her hairstyle, the Ponytail; it has become the icon for vintage dolls, the ponytail with curly bangs. Her face is very sophisticated and has a molded eyelash ridge. There are eight identifiable versions of this early doll with straight legs made from 1959-1964. The first three have turned a pale white; the first two have black and white — not blue — eyes and pointy eyebrows, and the first doll has holes in the ball of her feet and metal rods in her legs so she could stand on a plastic stand with two metal prongs. This first ponytail doll is called a #1 and can be sold for $4,000 if perfect. A played-with doll in excellent condition would be $2,000 or more.

138

#1 Barbie®, early 1959. The very first one.

Closeup of #1 Barbie®. (Note black and white eyes.)

uncommon for these dolls to be missing fingers and toes or have a neck split since little girls liked to swap their heads. In any condition, this doll could be priced at $1,000 because she is so very rare.

The #2 has no holes in her feet but is otherwise the same as the #1 and sells for about the same price. The #3 has blue eyes and softly curved eyebrows; she sells for $1,000. The rest, #4 to #8, are pink fleshed and have variations in their scalps, hip markings, hair texture, torso, etc. A careful examination of these dolls is needed to tell them apart. I recommend Sibyl DeWein or Kitturah Westenhouser's guides (See Bibliography). The #1 to #4 ponytails came in blonde or brunette only, but the #5 (1961) through #8 (1964) had various shades of the two colors plus red (Mattel, Inc. called this color Titian). These later ponytails are in the $100 to $400 range. A different ponytail style with a long lock of hair pulled from her front hairline to the ponytail knot was made in 1965 and called the Swirl, of course. She sells for $300.

Other hairstyles using this Ponytail face mold are the Bubble Cut, Fashion Queen, Bend Leg/ American Girl, and Color Magic. The Bubble Cut

Various Ponytails (left to right): #1 in 1959, #3 in 1960, #5 in 1961, #7 in 1963, and Swirl in 1964.

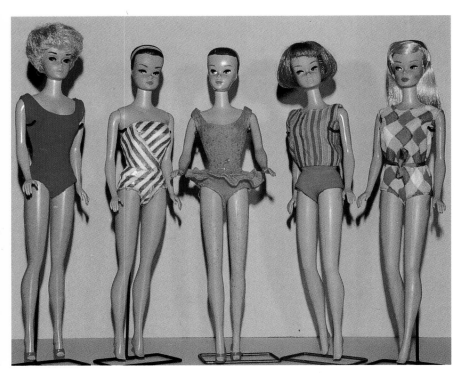

Other hair styles using the first face with molded eyelashes (left to right): Bubble Cut, Fashion Queen, Miss Barbie (a variation on this mold), Bend Leg, and Color Magic.

(BC) is a short, curly, and full style in many colors and is the most common vintage doll, probably because her popularity soared during her production years of 1961-1967. She is sold for $100 to $250, a light blonde with pale pink lips being the most expensive. Fashion Queen (1963-64) has molded brown hair (really bald) and came with three wigs that will melt to the doll's head over time if not wrapped in tissue or cloth. She has very heavy make-up. This doll sells for $100. The Bend Leg doll (B/L) began that feature in 1965 but is sometimes called the American Girl because she has a classic hairstyle of center part, bangs, and straight bob to her jawline. She and Color Magic, whose hair could change colors by applying the "secret formula," sell for $500. A special doll called Miss Barbie was made in 1964 only and actually was the first with bend legs. She has three wigs and a hollow head to accommodate open/close eyes, but her special head is so fragile that few survive and most have chunks missing with obvious repairs. Nevertheless, she is rare and sells for $300, more if she has her accessories: swing, plant, radio, and dishes.

All of these Ponytail dolls were made in Japan and most are marked as such. They are beautiful, high-quality dolls loved by collectors for their varying faces which were hand painted (and consequently easy to touchup, so buyer beware!). Their bodies are aging very well except for the Bend Legs, which develop splits. These doll fashions are the classics, copies of Parisian couture: full skirts or straight-skirted suits with hats, gloves, and matching handbags. We looked so good back then!

It is interesting to note that factory mistakes are highly prized by some collectors (though not all) and so these dolls may sell for exorbitant prices. For instance, a Bend Leg with her hair part sewn more to the side rather than centered might sell for $2,000. There is rampant piracy and dishonesty in altering all of these dolls because of their high prices, so be careful! Before buying, study the many books available and take along a knowledgeable friend.

The second face mold is my favorite, the Eyelash (1967 to 1976), but only the dolls through 1972 actually had individual rooted lashes. The face paint was softened so she appears much younger, more like a

Close-up of the Eyelash face mold, Talking Barbie® 1968.

Various hair styles using the Eyelash face mold (left to right): 1967 Twist N Turn, 1969 Twist N Turn (Flip), 1968 Talking, and 1971 Live Action.

teenager than with the sophisticated Ponytail look. These dolls have bend legs and pink flesh, but many of their faces are becoming lighter with age. The first year's issue had a new waist section that could move, thus her name Twist and Turn (TNT). She has bangs and long hair and sells for $150. Another TNT has a shoulder-length flip and one saucy forehead curl (1969 to 1971). This is my favorite because that's what I looked like then! Other versions include a doll with a talking mechanism activated by a pull string; the "living" bodies that are very wiggly with elbows and wrists that can be bent; and a doll with "growing" hair. These are in the $50 to $100 range but may soon go up in price because it's so hard to find a perfect doll: most have arm or leg problems. Still, collectors want them because they were designed to wear some of the most colorful and exciting outfits ever made by Mattel, Inc. Most of these dolls were made in Japan and are of very high quality like the Ponytails.

In late 1972 to 1976, the eyelashes were omitted. These dolls had no booklets to show the buyer what outfits were available and no tags in the clothes, though these things were included in all the dolls sold previously. Therefore they are less collectible because we didn't know who/what they were. These dolls and their clothing were made in various countries including Korea, Taiwan, and China because Mattel, Inc. was shopping the world market to keep manufacturing costs reasonable. They share the earlier dolls' arm/leg problems, but are still active dolls — some can hold things with a moveable thumb; others bow at the waist to play sports or have wiry hair that is easy to curl. They are lots of fun to pose, limb problems aside, and collectors will soon discover them.

The third face mold is the Superstar. She has a slight, toothy grin and permanently bent arms that move out from her shoulders. This mold has been the predominant one since 1977 and sells for less than $100. Please note again that many of these dolls are marked 1966 on their hips, but they are from much later. Mattel, Inc. used the old mold for decades and it fools lots of new collectors.

There were little plastic Barbie® doll ornaments and keychains, etc., big plastic dolls, a special line of

Close-up of Superstar face mold, 1977
to present. (Note teeth and bent arms.)

porcelain dolls, many more face molds for friends and family, and a stuffed body version at my last inventory. Mattel, Inc. is prolific; that's how they stay in business!

Friends, etc.

Ken®, the other friends, and the family dolls have never been as popular with collectors and so they don't command the prices that Barbie® dolls do. There are a few rare Skipper® and Francie® dolls—and some outfits considered rare—that sell for more than $100 each, but the majority of these dolls sell in the $10 to $50 range. This is good! This means there are still old dolls out there that are affordable! And any time a doll has his/her own line of clothing, it's a sure sign these will become specialized collectibles. For future reference, a good example of this is the Rockers(1986 and 1987): Six different dolls were made each year, one marketed only in Europe. These have their own line of outlandish and brightly colored clothing, musical instruments for all the members of the band, and a stage shaped like a guitar. This set is sure to be a favorite in another five to ten years.

Some of the Rockers (left to right): 1986 Barbie®, 1987 Ken®, and 1986 Diva®; $50 each. Scene: Rockers Cafe, 1986, Mattel, $70.

Chapter Eight

Barbie® Doll's boyfriend has been **Ken**® doll since 1961, except for 1967 and 1968, two years when he mysteriously disappeared. (Mine dated GI Joe by Hasbro, Inc. for the interim.) Ken® doll was back by 1969. They have never married—except a million times at your house and mine—though the bridal dress is always the best-selling fashion of any year's line. We are truly romantics! His first rooted hairstyle (brown) came on Mod Hair Ken® doll in1973, his first earring in 1993.

Midge was Barbie's first girlfriend, from 1963 to 1967, then went on a long vacation until she was reintroduced in 1988. She is a favorite with collectors because she has freckles. Her boyfriend from 1964-66, Allan, also has freckles.

Skipper, the little sister, has not been out of production since she was introduced in 1964. She has had many friends — both male and female — and a line of clothing smaller in number but just as beautiful as her older sister's. Some of their outfits match, and these are the most popular with collectors, of course. We don't know how old she really is, just younger than her sister, but she grew a bust in 1975! That was the year she started growing — no joke. This doll gets a little chesty and a tiny bit taller when her arm is rotated! She even has a friend, Ginger, who does the same thing. This doll was the ultimate adolescent fantasy and was probably only bought by liberal moms of the '70s. Today, these dolls are hoarded by every self-respecting Skipper collector.

Barbie® doll **hair colors** used to be blonde, brunette, and red. She became a platinum blonde exclusively in 1971. Friends were created that had differing hair and facial features to represent various ethnic groups. That changed in 1980 with a Black and Hispanic version of the Barbie® doll followed by an Oriental in 1981 (another series that has become a specialty collector area). The pink-flesh Barbie® was once again manufactured in brunette and redhead as Teen Talk of 1992. She was also the first talking doll since 1973.

In 1995, **Mattel, Inc.** reached a **production** proliferation landmark by offering more than a hundred different dolls which were named Barbie® doll, her friends or family. Many of these were made especially for various department stores (DSS) and so are difficult to find. This does not make these dolls expensive, however, and some DSS dolls are costumed beautifully. J.C. Penney's "Enchanted Evening Barbie" (1992) is an earlier example with a diamond-pattern, multi-colored, metallic full-length coat trimmed in white fur with matching hat and purple metallic evening dress; she is stunning. Mattel's productivity and the quality of competitors' dolls have made it necessary for most collectors to specialize. There are Vintage collectors (1959-72 with a sub-specialty of Mod collectors (1967-72), Pink Box collectors (1977 to present), those who buy only DSS Exclusives and Limited Editions (a new Mattel marketing strategy), specialized doll collectors who only buy dolls like Skipper or Francie, and those who are like me and buy a little of everything I like. There are doll shows to sell you anything you miss and doll museums to visit if your house can't hold them all!

Don't be sad if you missed the vintage dolls in the '60s and want them now but find them too expensive. Mattel, Inc. has been reissuing a few of these **nostalgia dolls in retro outfits** each year since 1994. They are not exactly like the older ones, but they are close enough to be lots of fun for any collector to own.

In 1990, Mattel, Inc. began to market a special doll line designed by the world famous **Bob Mackie**. His early costume designs for such stars as Carol Burnett and Debbie Reynolds made him a huge success.In the late 1970s he created doll fashions for Mego's Cher, Tenille, Farrah, and Diana Ross dolls which have become highly collectible. His Mattel doll designs are the ultimate in glamor and beauty, with several now selling for near $1,000 (they were originally $150 to $200 with doll).

If you **clean old dolls** with anything stronger than soap and water, you are risking severe damage. Cleaners may react with the plastics to form powders and crusts, and other unfortunate results can occur such as bleached areas, spots resembling bruises, etc. Green ears from earrings are a big problem in vintage dolls, but no reliable remover is known and the whole head could end up green (or bleached out). The same is true of the clothes. Some are cotton and can be laundered, but even these tend to shrink and they will lose their value if cleaned. Most are of other fabrics (silk, wool, taffeta) and have become extremely delicate over time. Repair with care! It's best to leave them alone and let them age gracefully.

The early **outfits were tagged** with a distinctive Barbie® doll label woven in black and white threads. The earliest were labeled "TM;" the later ones with an "®" mark. These labels were found in only *one* clothing piece per outfit, however, usually the upper and outer piece. Thus a skirt or slacks might not have a tag (unless sold separately as a Pak item). Ken® doll, Skipper, Francie, and Tutti had their own clothing labels, as did Julia, but the other friends did not. The tags were omitted beginning in 1972 with the "Best Buy" line and did not appear again until 1977. These newer tags say "Genuine Barbie®" doll and appear to be printed, not woven. The two tags are easy to tell apart. These tags are not numbered by year. A collector needs books to identify outfit pieces and era. The best books are Sarah Eames' *Barbie Doll Fashion, Vols. I and II* (1959 to 1974) and my books, *The Eyelash Era Fashions (1967-1972)* and *Ken Fashions (1961-1976)*.

The **copyright date** on Mattel, Inc. boxes and other products is one year *before* it was actually available on the store shelves. This creates mayhem for collectors in dating items: which date to use?

The **year on the doll's hip** is not necesarily the doll's issue date. It's the mold copyright year, and molds are reused, sometimes for decades.

Bryan, Sandra. *Barbie, The Eyelash Era Fashions 1967-1972*. P.O. Box 162562, Austin, 78716, 1989.

This was the first research published showing each of these outfits. Pocket size with black and white photos.

Bryan, Sandra. *Ken Fashions 1961-1976*. P.O. Box 162562, Austin, 78716, 1990.

This was the first published research showing each piece of these outfits and the first to include the 1972 to 1976 lines. Pocket size with black and white photos.

DeWein, Sibyl, and Joan Ashabraner. *Collectors Encyclopedia of Barbie Dolls and Collectibles*. Paducah, KY: Collector Books, 1977.

This was the first research on the dolls, clothes, and accessories published. It is the best source for identifying the early dolls. Most photos in black and white.

Eames, Sarah Sink. *Barbie Doll Fashion*, Vols. I and II. Paducah, Kentucky: Collector Books, 1990 and 1996.

These are color photos of every outfit for every doll through 1974.

Manos, Paris and Susan. *The World of Barbie Dolls*. Paducah, KY: Collector Books, 1983.

The first pocket-size book in color featuring all the dolls through 1978 and many outfits.

Melillo, Marcie. *The Ultimate Barbie Doll Book*. Iola, WI: Krause Pub., 1996.

This color reference describes every doll in its package from 1959 to 1995, with close-ups of each face.

Miller, Barbara and Dan. *Miller's Price Guide Pocket Annual*. P.O. Box 8722, Spokane, 99203.

This guide lists prices for vintage and new Barbie® dolls, clothes, etc. They also publish a monthly market report and quarterly magazine with very current information.

Westenhouser, Kitturah. *The Story of Barbie™*. Paducah, KY: Collector Books, 1994.

Not as detailed as the DeWein text, but still an excellent identification guide and overview for collectors. All color and includes the more current dolls.

Your Own Barbie Scenes

Notes:

_____   _____
_____   _____
_____   _____
_____   _____
_____   _____
_____   _____
_____   _____
_____   _____
_____   _____
_____   _____
_____   _____
_____   _____
_____   _____

154